LIMITATIONS OF HUMAN MIND

That Restrain us From Visualizing the Existence of God and Souls

By

SUBHASH CHANDRA SAWHNEY

Copyright © 2024 by Subhash Chandra Sawhney

All rights reserved. No part of this book may be reproduced in any form without permission in writing from the author.

No part of this publication may be reproduced or transmitted in any form or by any means, mechanical or electronic, including photocopying or recording, by any information storage and retrieval system, or by email or any other means whatsoever without permission in writing from the author.

THE REASON WHY YOU NEED TO READ THIS BOOK

You need to read this book because besides furnishing some such proofs of the existence of God and souls that supersede all such proofs that have failed to convince all those who have started doubting whether the entities known as God and souls exist – it also tells about two such interesting things that are true even though nobody seems to be aware of them, to drive home the idea that it is wrong for them to think that God and souls may not exist simply because they are not aware of their existence.

PROLOGUE

This book begins with a chapter that tells about four such occasions when science admitted such things to be true, which it has been all along, treating not to be true in the past with a hope that it should not come to us as a big surprise if, some day — it may also admit that the entities known as "God" and "souls", too, exist even though it is very repugnant to believe it to be true.

In the second chapter of the book, I have told that our brain is not able to know the answers to the following type of questions only in the same way as *"our eyes are not able to see the E/M waves of frequencies below the frequency of red waves and the E/M waves of frequencies above the frequency of violet waves"* even though such waves exist and *"our ears are not able to hear the waves of the frequencies below 20 Hz and the waves of the frequencies above 20,000 Hz"* even though waves of such frequencies also exist.

- Though the riddle "If nothing should have existed before the Big Bang, how the type of energy supposed to have existed before the Big Bang should have come into existence?" has not been solved until now; it may be solved by assuming that God may have created it, we can't answer — if so, how God should have come into existence?"

- If God should have known the technique of producing such energy in so much quantity as should have existed, from nothing — 'Why we don't know how to produce anything from nothing?"

- How He would look like if we could see Him?

The next two chapters of this book tell about such things that are true even though nobody seems to be aware of them — to push forward the argument that it is wrong to think that God may not exist just because nobody is aware of its existence.

The next two chapters tell about two such "Information Systems" which could have been developed only by some such entity that may have had a brain thousands of times more capable than the human brain to push forward the argument that we should not mind treating such superhuman entity as "God".

This argument supersedes all other arguments advanced by others in this context — until now.

While the seventh chapter of the book tells about the mathematical reason which proves that it should be possible to produce every type of thing from nothing — the next five chapters of the book, tell about such things which can't be explained unless we assume that some invisible entities known as "souls" exist even though they do not exist physically exactly in the same way as certain astronomical observations can't be explained without assuming that the invisible things known as "dark matter" and "dark energy" exist— to push forward the argument that since science has acknowledged the existence of invisible "dark matter" and "dark energy" even though such things do not physically exist; it should not be reluctant to admit the existence of the souls also, simply because they are neither visible nor do they exist — physically.

TABLE OF CONTENTS

FOUR OCCASIONS WHEN SCIENCE BROUGHT EVEN SUCH THINGS WITHIN THE AMBIT OF SCIENCE THAT DID NOT FORM A PART OF SCIENCE IN THE PAST 11

LIMITATIONS OF HUMAN MIND 15

SOMETHING THAT IS TRUE EVEN THOUGH NOBODY IS AWARE OF IT 21

ONE MORE SUCH THING THAT IS ALSO TRUE EVEN THOUGH NOBODY IS AWARE OF IT 29

VISUALIZATION OF A PROOF OF THE EXISTENCE OF GOD 37

VISUALIZATION OF ANOTHER PROOF OF THE EXISTENCE OF GOD 51

THE WAY WE MAY PROVE THAT IT SHOULD BE POSSIBLE TO PRODUCE EVERY TYPE OF THING FROM NOTHING 57

RECENT EVIDENCE THAT SUPPORTS THE "THEORY OF REINCARNATION" 61

PIECES OF EVIDENCE OF REINCARNATION — COLLECTED BY DR IAN STEVENSON 65

THE TYPE OF FUNCTIONS ONLY SOULS MAY BE ABLE TO PERFORM 71

IT LOOKS AS THOUGH THE CURSES AND BLESSINGS GIVEN BY US ARE EXECUTED BY OUR SOULS ONLY ... 77

HOW SOULS MAY BE ASCERTAINING WHO HAS TO GET REINCARNATED WHEN AND WHERE............................ 87

THE WAY HINDU DEITIES PROVE THEIR EXISTENCE BY APPEARING IN THE FORM OF THEIR APPARITIONS 91

A FACT THAT MAY DISAPPOINT THOSE WHO THINK — GOD MAY BE OMNIPOTENT.. 99

THE REASON WHY WE DON'T KNOW SOME OF SUCH TECHNIQUES THAT WERE KNOWN TO SOME PEOPLE IN THE PAST 105

PARA SCIENCE RELATED TO THE EPISTLE WRITTEN BY PAUL IN CHRISTIAN NEW TESTAMENT 117

DISCLAIMER 123

PROFILE OF THE AUTHOR 125

MAY I ASK YOU FOR A SMALL FAVOR? 127

FRONT COVERS OF THE OTHER BOOKS OF THE SAME AUTHOR 129

CHAPTER - ONE

FOUR OCCASIONS WHEN SCIENCE BROUGHT EVEN SUCH THINGS WITHIN THE AMBIT OF SCIENCE THAT DID NOT FORM A PART OF SCIENCE IN THE PAST

Though there may be even a few more occasions when science may even such things within the ambit of science that did not form a part of science in the past— I am, presently, aware of only the following four such occasions.

FIRST OCCASION

Though, later on, it could become possible to prove that negative numbers may have even real cube roots and real sixth roots; the fact that "negative numbers may also have real square roots" had been recognized to be true only when Girolamo Cardano told the world about his discovery that the square roots of the equation "$x^2 - 10x + 40 = 0$" incorporate the square roots of the negative number "-15", in his Latin-language book Ars Magna (The Great Art) in the year 1545 that proved that negative numbers may also have real square roots though in the past — everybody thought that negative numbers may have only imaginary square roots.

SECOND OCCASION

Though now, we know that even the leaves of all carnivorous plants possess consciousness — science was made aware of the fact that just like humans, plants too have feelings by placing a plant inside a vessel containing bromide solution, which is poisonous by Jagdish Chandra Bose when he demonstrated in the year 1901 at the Royal Society of London using his instrument to show on a screen — how the plant responded to the poison.

THIRD OCCASION

When in the year 1933, it was discovered that some invisible thing known as "dark matter"exists even though it does not physically exist since such matter was needed to explain the existence of fast-moving galaxies in the Coma Cluster.

FOURTH OCCASION

When in the year 1988, two different teams of astronomers led by astronomers Adam Reiss, Saul Perlmutter, and Brian Schmidt observed that the fact that at a certain redshift, the stellar explosions were dimmer than expected— may not be explained unless we assumed that some energy such as "dark energy" exists which is not only invisible, it also does not exist physically, for which they were awarded the 2011 Nobel Prize in Physics.

Don't you think — science should tender an apology for having dissuaded people from believing in the existence of God and souls for wrong reasons because it has retracted from such criteria by having acknowledged the existence of "dark matter" and "dark energy" even though both of them are neither visible nor do they physically exist?

Especially so — because in the same manner in which the existence of "dark matter" and "dark energy" has been justified; the existence of God and soul may be also justified exactly in the same manner, as has been elaborated by me in this book.

CHAPTER - TWO

LIMITATIONS OF HUMAN MIND

Is it not true that though we know that all the matter we observe today could not have come into existence unless space should have been filled with a type of unstable form of energy which should have transformed at some instant into the fundamental particles of which all types of matter are made of, during Big Bang — we don't know till today, from where such energy should have arrived?

Don't you think, if nothing should have existed before such unstable type of energy — some such an entity should have existed which should have known the technique of producing such energy from nothing?

SINCE WE ARE NOT ABLE TO PRODUCE ANYTHING FROM NOTHING; DON'T YOU THINK — WE SHOULD NOT BE AVERSE TO REVERE SUCH AN ENTITY AS "GOD" THOUGH IT BEWILDERS US, "HOW SUCH AN ENTITY SHOULD HAVE COME INTO EXISTENCE?"?

The fact is, I have spent a lot of my time thinking "*Why we are unable to know how such an entity could have come into existence?*" and finally — the following reason has struck me.

The reason why we not only don't know how the entity known as "God" should have come into existence — we may never be able to know how such an entity should have come into existence.

Nor may we ever be able to know — how He may look like.

The reason may be, very well, understood with the help of the following diagram of the spectrum of electromagnetic waves.

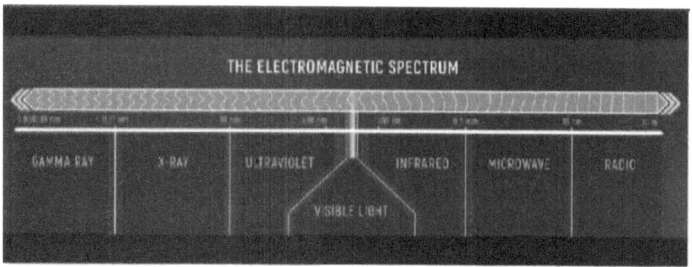

Is it not true that we know that our eyes can see only a small segment of the spectrum of the electromagnetic waves represented by the portion of the spectrum lying between 400 THz and 800 THz, even though the electromagnetic waves of lower wavelengths as well as higher wavelengths also exist?

Likewise, is it also not true that we know that our ears can hear only a limited range of the frequencies of sound, even though the frequencies corresponding to the ultrasonic and the infrasonic waves also exist?

But if so, does it not surprise you — why it should have not occurred to anyone until now that what is true about our eyes and ears, may be true for our brain also?

SO SHOULDN'T WE AGREE THAT, IN THE SAME WAY, EVEN OUR BRAIN MAY BE ABLE TO UNDERSTAND ONLY A LIMITED TYPE OF THINGS — NOT EVERY TYPE OF THING?

I hope, it explains not only the reason, why we don't know "How such entity could have come into existence?" but even the reason — why we do not know some of such things, as

- How to produce anything from nothing though such entity should have known the techniques of producing all types of things from nothing?

- Though no offspring may get born until and unless a male creature and a female creature cohabit — how the first couple of each type of creature living on the Earth, should have come into existence? Though we don't know — how such couples should have come into existence, don't you think — some entity should have known the technique of creating such couples since if such couples had not come into existence — all such creatures could have not come into existence?

- In the same way, since no hen may lay an egg unless a cock may mate with it; some entity should have known the technique of creating the first specimens of a cock and a hen also.

- Who could have developed the type of information systems described by me in the third and fourth chapters of this book?

- Where do the souls reside after departing from the body of their previous host/hostess till they get reincarnated?

- What type of astrology the souls may be using to ascertain when and where their previous host/hostess should get reincarnated?

Of course, this is not the only reason why we don't know many of such things.

There is one more reason why we don't know many of such things.

Such reason has been explained by me in the next two chapters.

CHAPTER - THREE

SOMETHING THAT IS TRUE EVEN THOUGH NOBODY IS AWARE OF IT

We may prove the existence of God only after realizing that it is not necessary that only such things may be true of which, we may be aware in the same way as the following thing is true though nobody is aware of it.

Just think — how many of us know that all of us are travelling in space at a speed as high as 1027.7 times the speed of sound, though it is true that all of us are travelling in space at a speed as high as 1027.7 times the speed?

If you think, I may be joking — no, I am not joking.

Would you believe — I have already told the method of calculating the speed at which all of us are traveling in space in the article "**Whether you know or not — all of us are travelling in space at a speed as high as 1027.7 times the speed of sound**", which had been posted by me on medium.com, on October 9, 2023? [1]

I am going to reproduce this article here only with the intention of making you aware of the fact that much in the same way, as it is true that all of us are travelling at such a high speed in space even though we are not aware of it — it is wrong to think that we should not agree that some superhuman entity known as God may exist just because we are not aware of its existence.

We may calculate the speed at which we are travelling in space, as follows.

The manner in which we may calculate the speed at which all of us are travelling in space

Is it not true that we also revolve around the Sun at the same speed at which the Earth revolves around the Sun?

Well, since the Earth revolves around the Sun at an average speed of about 1,674 km/hr — it implies that we also revolve around the Sun, at such speed.

Any doubt about it?

Next ...

Though we are travelling at this speed around the Sun, the fact is — the Sun is revolving around the black hole of the Milky Way also along with its all planets (including the Earth) at an average speed of about 828,000 km/hr.

So we are also travelling around the black hole of the Milky Way at the same speed.

However, because the direction in which the Earth is moving around the Sun and the direction in which the Sun is moving around the black hole are not the same, we are not travelling at a speed of 1,674 + 828,000 km/hr — the arithmetic sum of their speeds.

We have to treat the speed 1,674 km/hr as well as the speed 828,000 km/hr as vectors.

If we may treat the speed 1,674 km/hr as the vector "$\vec{V_1}$" and the speed 828,000 km/hr as the vector "$\vec{V_2}$", we may say — we are travelling in space at a speed of "$\vec{V_1} + \vec{V_2}$", as shown in the following diagram.

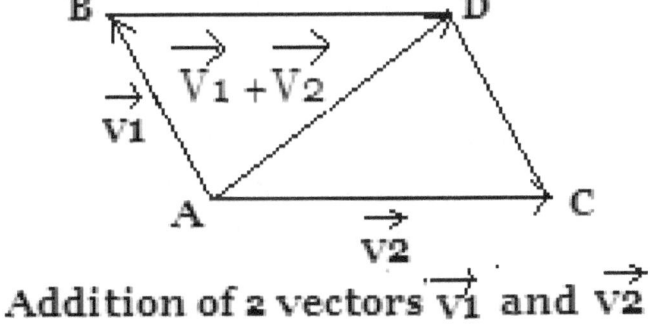

Addition of 2 vectors $\vec{V_1}$ and $\vec{V_2}$

Though we should calculate the value of "$\vec{V_1} + \vec{V_2}$" using the formula "$\vec{V_1} + \vec{V_2} = \sqrt{(V_1^2 + V_2^2 + 2\, V_1.V_2 \cos A)}$" where ∠A is ∠BAC, the fact is — we can calculate the value of "$\vec{V_1} + \vec{V_2}$" by this method only if we know the value of "∠A".

Since the value of "∠A" keeps on changing; obviously, this method does not suit us.

So we need some shortcut.

Luckily, we don't have to hunt for a shortcut.

I have found out the shortcut.

THE SHORTCUT

We may calculate an approximate value of "$\vec{V_1} + \vec{V_2}$" even by multiplying "V1 + V2" by "0.9" since such value is expected to lie between 0.85 to 0.95 times the value of "$\vec{V_1} + \vec{V_2}$", most of the time.

So we may say, we are travelling at a speed of approximately 0.9 times "828,000 + 1,674" that is — at a speed of about "746,706.6 kilometers per hour" in space.

But the fact is, the Sun is not travelling only around the black hole.

It also oscillates as shown in the following diagram.

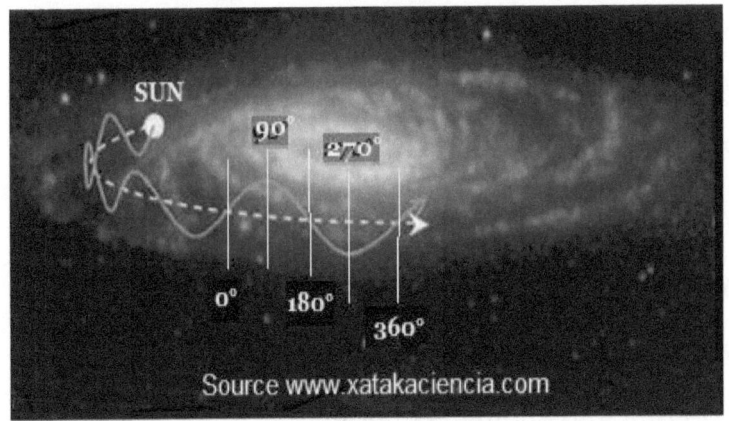

Source www.xatakaciencia.com

At present, it is climbing at a speed of approximately 25,200 kilometers per hour above the galactic plane.

Though you shall get a different value of this speed depending upon the source from where you may pick up this figure, I picked up this figure from "Poe".

If we may denote this speed of the Sun by the vector "$\vec{V_3}$" and add it to "$\vec{V_1} + \vec{V_2}$" by multiplying "746,706.6 + 25,200" by "0.9", we shall get the approximate value of "$\vec{V_1} + \vec{V_2} + \vec{V_3}$" which works out to a figure of "694715.94 kilometers per hour".

But besides climbing above the galactic plane, Sun is also travelling toward the constellation Hydra along with its all planets at an approximate speed of 72,000 kilometers per hour.

Let us designate this speed as the vector "$\vec{V4}$", add it to "$\vec{V1} + \vec{V2} + \vec{V3}$" and multiply it by "0.9" to arrive at a value of "$\vec{V1} + \vec{V2} + \vec{V3} + \vec{V4}$".

If we multiply "694715.94+72,000" by "0.9", it yields a figure of "690,044.346 kilometers per hour".

So we may say, the value of "$\vec{V1} + \vec{V2} + \vec{V3} + \vec{V4}$" is around "690,044.346 kilometers per hour".

According to "Poe", our galaxy is moving toward the Centaurus constellation at an approximate speed of 720,000 km/hr.

If we may designate this speed by the vector "$\vec{V5}$", add it to "$\vec{V1} + \vec{V2} + \vec{V3} + \vec{V4}$" and multiply it by "0.9", we may calculate the value of "$\vec{V1} + \vec{V2} + \vec{V3} + \vec{V4} + \vec{V5}$".

If we multiply "690,044.346 +720,000" by "0.9", it yields a figure of "1,269,039.9 kilometers per hour".

SO WE MAY SAY, THE VALUE OF "$\vec{V1} + \vec{V2} + \vec{V3} + \vec{V4} + \vec{V5}$" IS ABOUT "1,269,039.9 KILOMETERS PER HOUR", WHICH IMPLIES THAT WE ARE CONSTANTLY TRAVELLING IN SPACE AT AN AVERAGE SPEED OF ABOUT 1027.7 TIMES THE

SPEED OF SOUND BECAUSE THE SPEED OF SOUND IS 1234.8 KM/HR — ANY DOUBT ABOUT IT?

---

[1] https://subhashchandrasawhney.medium.com/whether-you-know-or-not-all-of-us-are-travelling-in-space-at-a-speed-as-high-as-1027-7-eccec512eff4ot

CHAPTER - FOUR

ONE MORE SUCH THING THAT IS ALSO TRUE EVEN THOUGH NOBODY IS AWARE OF IT

Much in the same way, as the stewardess of an airplane tells us to fasten our seatbelt when the plane in which we may be travelling may be going to descend, I may also tell you to fasten your seatbelt before I tell you how we may calculate the number of the dimensions of the path traced by any satellite of any planet in space again with the purpose of suggesting that it is quite possible that we don't know — whether God exists or not, in the same way as we don't know that the satellites of all planets trace a multidimensional path in space.

Well, once I have driven home the purpose of my telling, I don't mind telling the way — we may calculate the number of dimensions of the path the satellites of various planets trace in space.

This is how we may calculate the number of the dimensions of the path the satellites of all planets trace in space.

THE NUMBER OF THE DIMENSIONS OF THE PATH THE SATELLITES OF ALL PLANETS TRACE IN SPACE

Though we can't see such dimensions because our eyes may see only three dimensions; it is true that we may visualize that all satellites of all planets trace a multidimensional path in space.

Dimensions which we can't see also exist much in the same way as besides the visible part of the spectrum of the E/M waves, the parts of the spectrum extending beyond the visible part of the spectrum — also exist.

The only difference is, though we can't see them — we can very well, perceive them in such a way as I had explained in the answer given by me to a query related to the higher dimensions raised by someone on quora.com. [1]

THE MANNER IN WHICH WE MAY VISUALIZE THE EXISTENCE OF INVISIBLE DIMENSIONS

I am reproducing the answer given by me to this query, below.

I shall tell you how we may visualize the existence of four dimensions as well as not only five dimensions but even more than five dimensions .

It is only the distance from where we see an object that makes all the difference.

It is only because we see most of the things from a pretty close distance that we can see even the depth of an object besides its width and height.

For instance, we can't see the depth of the moon when we look at it from the Earth.

Doesn't it appear to us as a two-dimensional object even though we pretty well know – it is a three-dimensional object?

So it explains that we can't tell how many dimensions an object may have simply based on what we can see through our eyes.

We may explain the fact that all satellites of all planets trace a multidimensional path in \space through the following example.

When we look at the moon from the Earth, it appears to be moving in a two-dimensional elliptical orbit around the Earth, to us.

It is so because while standing on the Earth, we are neither able to see the Earth revolving around its axis nor whether it is revolving even around the sun.

Just imagine how it would look like if we moved out to a point from where we could see both motions of the Earth.

From such a point, it would appear to be moving as if it is moving along a three-dimensional cylindrical helical path, as shown below.

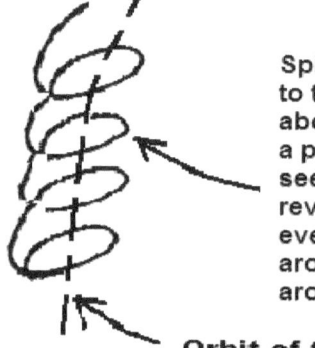

Spiral path the moon appears to trace around the earth in about 4 days when viewed from a point from where we could see not only the moon revolving around the earth but even the earth revolving around its own axis as well as around the Sun

Orbit of the earth

Distance covered by the earth around the Sun in 4 days = 1,02,95,080 km
Rounds taken by the earth around its own axis in 4 days = 4

The moon, of course, continues to have three dimensions only.

A solid object like moon may never have more than three geometrical dimensions – "width", "depth" and "height".

Only the path traced by it in the space acquires more than three dimensions.

Moon still has three dimensions only.

Now let us extend this logic further.

As you know, Earth even wobbles around its axis as shown below.

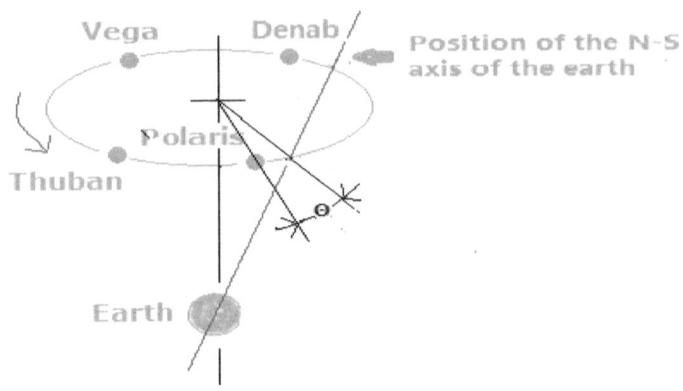

Thuban pinpointed the north celestial pole in the year 2787 BC and Kochab served as a pole star in 1100 BC.

Of course, since it wobbles once in about 26000 years we would not be able to see that it is wobbling. So, it would not make the helical path look like a four-dimensional object.

But as you know, Sun oscillates above and below the galactic plane as well.

Imagine yourself looking at the moon from a point from where you could see the Earth oscillating along with the sun, as shown below.

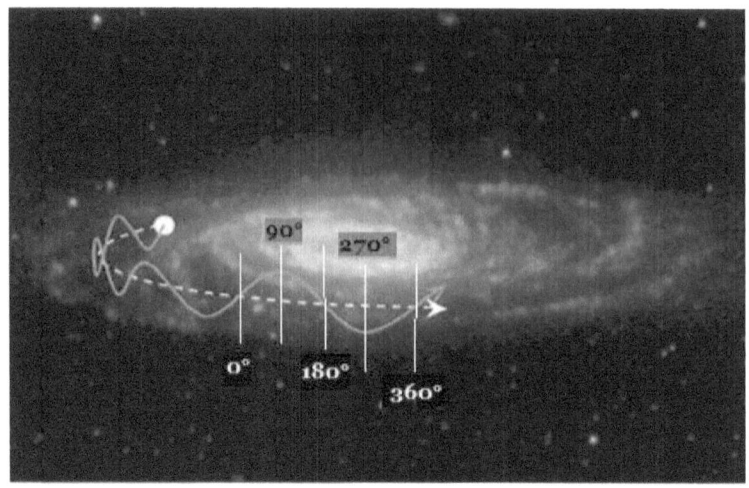

Don't you think - the same helical path should now acquire a four-dimensional look?

It is quite another thing though whether you can see the moon from that far a distance or not. Surely, we can't say that we can't see it even through a telescope from such a point

So we can't deny that the path traced by the moon would not look like a four-dimensional object.

But we also know that the sun revolves even around the centre of the galaxy.

If you could see the moon also revolving along with the Earth and the sun — around the centre of the galaxy, should you doubt that the same helical path would not take a five-dimensional shape?

The fact is — the same path which looks like a 5-D path if we look at it from a place from where we may be able to see the moon also revolving along with the Earth and the sun — around the centre of the galaxy.

Though I had told about the path traced by the moon only to the extent of 5-D; we may extend the same logic to the level of even 7-D, in the following manner.

Is it not true that the Sun is not travelling only around the black hole — it is also travelling toward the constellation Hydra along with its all planets and their satellites?

So do you doubt, if we may look at the path traced by the moon from a place from where it may be possible to see the Sun travelling toward the constellation Hydra — it should look like a 6-D path from there?

Going a step further — is it also not true that even our galaxy is moving toward the Centaurus constellation?

So don't you think — if we could look at the path traced by the moon from such a place from where we may be able to see even our galaxy travelling toward the Centaurus constellation — the path traced by the moon in space should look like a 7-D path from there?

BUT JUST THINK — WHAT THE PREVIOUS TWO CHAPTERS AND THIS CHAPTER TELL.

These three chapters tell that though it may not be possible for us to know how God should have come into existence and why we may never be able to know how to produce anything from nothing — it should be possible for us to know whether God exists or not in the same way as we can know that all of us are travelling in space at a speed as high as almost 1027.7 times the speed of sound, as well as to know that it is possible to visualize that satellites of all planets trace a multidimensional path in space.

The next two chapters of the book enable us to know, whether any super-human entity having a brain thousands times more capable than the human brain exists or not and if it exists — why should we mind treating such entity as "God"?

---

[1]   https://www.quora.com/How-do-one-can-feel-more-than-three-dimensions/answer/Subhash-Chandra-Sawhney — posted by me on quora.com on February 8, 2018

CHAPTER - FIVE

VISUALIZATION OF A PROOF OF THE EXISTENCE OF GOD

This chapter of the book lays bare the fact that the human anatomy also consists of two parts — an visible part and an invisible part.

Though Gray's book of anatomy tells about the visible part of the human anatomy in detail; it does not tell anything about the invisible part of human anatomy.

Since God is an invisible entity, its evidence can't be understood by understanding the visible part of the human anatomy.

It may be understood only by understanding the invisible part of human anatomy — not the visible part of human anatomy.

I am going to make you aware of the invisible part of the human anatomy through this chapter of the book after briefly touching upon the visible part of the human anatomy.

Though the entire medical faculty knows about the sequence in which a human fetus grows from the date of the commencement of the pregnancy onward week by week — I want to acquaint the non-medicos also about such sequence which forms a part of visible anatomy.

If you are a medico, you may, very well, skip this part and go, straightaway, to the next part of the chapter which tells about the invisible part of human anatomy.

THE VISIBLE PART OF HUMAN ANATOMY

It tells that various parts of the human fetus grow in the following sequence during pregnancy.

(i) During the first week and the second week of the pregnancy

From the day a woman starts menstruating, no woman may become pregnant till the end of the second week because an egg is released from her ovary only during the end of the second week.

A woman conceives when, out of all the sperms that enter her vagina during copulation — the strongest sperm travels through the cervix (the opening of the womb, or uterus) toward the fallopian tube and meets the woman's egg cell as shown in the following diagram resulting in the formation of a cell known as "zygote".

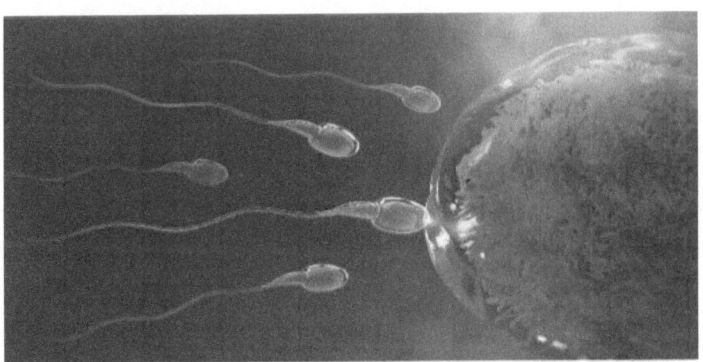

(ii) During the third week of the pregnancy

Thereafter, the zygote spends the next few days traveling down the fallopian tube. During this time, it divides to form a ball of cells called "blastocyst", which is made up of an inner group of cells with an outer shell.

While the inner group of the cells will become the embryo — the outer group of cells will become membranes, which nourish and protect the embryo.

(iii) During the fourth week of the pregnancy

During the fourth week, the "blastocyst" reaches the uterus, it buries itself in the uterine wall.

At this point in the woman's menstrual cycle, the lining of the uterus is thick with blood and ready to support a baby.

(iv) During the fifth week of the pregnancy

The "embryonic period" starts from the fifth week onward.

This is when all the baby's major systems and structures develop. The embryo's cells multiply and start to take on specific functions.

During the fifth week, blood cells, kidney cells, and nerve cells develop.

The embryo grows rapidly, and the baby's external features begin to form.

The baby's brain, spinal cord, and heart begin to develop.

The baby's gastrointestinal tract starts to form.

(v) During the sixth week and the seventh week of the pregnancy

During the sixth and seventh weeks, arm and leg buds start to grow and the baby's brain forms into 5 different areas.

Eyes and ears begin to form.

Tissues grow that will become the baby's spine and other bones.

The baby's heart continues to grow and starts beating at a regular rhythm.

(vi) During the eighth week of the pregnancy

During the eighth week, the baby's arms and legs grow longer.

Hands and feet begin to form and look like little paddles.

The baby's brain continues to grow and the lungs start to form.

(vii) During the ninth week of the pregnancy

During the ninth week, nipples and hair follicles form.

Arms grow and elbows develop and the baby's toes can be seen.

(viii) During the tenth week of the pregnancy

During the tenth week, the baby's eyelids get developed and begin to close. The outer ears begin to take shape. The baby's facial features become more distinct and the intestines begin to rotate and at the end of the 10th week of pregnancy, the baby is no longer an embryo. It is now a fetus.

(ix) During the eleventh week up to the fourteenth week of the pregnancy

During the eleventh to fourteenth weeks, the baby's eyelids close and will not reopen until about the 28th week. The baby's face gets formed. The limbs become long and thin and nails appear on the fingers and toes. Genitals appear and the baby's liver starts making red blood cells. The baby can now make a fist and the tooth buds appear for the baby teeth.

(x) During the fifteenth week up to the eighteenth week of the pregnancy

During the fifteenth to eighteenth weeks, the baby's skin is almost transparent. Fine hair called lanugo develop on the baby's head. Muscle tissue and bones keep developing, and bones become harder and the baby begins to move and stretch. The liver and pancreas start producing secretions.

(xi) During the nineteenth week up to the twenty-first week of the pregnancy

During the nineteenth to twenty-first weeks, the baby is able to hear. The baby becomes more active and continues to move and float around.

(xii) During the twenty-second week of the pregnancy

During the twenty-second week, Lanugo hair covers the baby's entire body. Meconium, the baby's first bowel movement, is made in the intestinal tract. Eyebrows and lashes appear.

The woman can feel the baby moving and the baby's heartbeat can be heard with a stethoscope.

The nails grow to the end of the baby's fingers.

(xiii) During the twenty-third week up to the twenty-fifth week of the pregnancy

During the twenty-third to twenty-fifth weeks, bone marrow begins to make blood cells.

(xiv) During the twenty-sixth week of the pregnancy

During twenty-sixth week, eyebrows and eyelashes get well-formed. All parts of the baby's eyes get developed and air sacs get formed in the baby's lungs.

(xv) During the twenty-seventh week up to the thirtieth week of the pregnancy

During the twenty-seventh to thirtieth weeks, the baby's brain grows rapidly. The nervous system is developed enough to control some body functions. The baby's eyelids can open and close. The respiratory system, while immature, produces surfactant which helps the air sacs fill with air.

(xvi) During the thirty-first week up to the thirty-fourth week of the pregnancy

During the thirty-first to thirty-fourth weeks, the baby grows quickly and gains a lot of fat. Rhythmic breathing occurs though the baby's lungs are not fully mature and the baby's bones get fully developed.

(xvii) During the thirty-fifth week up to the thirty-seventh week of the pregnancy

During the thirty-fifth to thirty-seventh weeks, the baby weighs about 2.5 kilograms and keeps gaining weight, but will probably not get much longer. The skin is not as wrinkled as fat forms under the skin and the baby has definite sleeping patterns.

(xviii) During the thirty-eighth week up to the fortieth week of the pregnancy

During the thirty-eighth to fortieth weeks, lanugo disappears except for on the upper arms and shoulders. The fingernails may extend beyond the fingertips. Small breast buds are present in both sexes. Head hair is now coarse and thicker and, now, a baby may be born any day.

Fine!

Now let me tell about the invisible part of human anatomy.

THE INVISIBLE PART OF HUMAN ANATOMY

This part of anatomy tells that a fetus could have not grown at all unless some entity should have developed the type of information system which may be telling each cell "What it is supposed to do" and "How it may do what it is supposed to do".

The purpose of telling you about such system is to convince you that since such system could have been developed only by some such entity that should have had a brain thousands times more capable than the human brain, we should not mind calling any such entity — "God".

Science could not delve on this part of anatomy because it called it a day after discovering such sequence as has been described by me just now.

It could have discovered that it would have not been possible for the fetus to grow in such sequence unless some entity ought to have developed the following type of information system — only if it would have not called it a day after having worked on the visible part of human anatomy.

Though science has been able to find that by the end of the fortieth week, the fetus consists of about 38 trillion cells of about 300 types; science did not bother to think, how all the cells produced from the day the zygote got formed — could have come to know the following type of things during the entire period of gestation.

(i) How many cells of each type have to be produced by them

(ii) How each type of cell to be produced by them — may be produced

(iii) Geometrically where exactly, they have to locate each baby cell produced by them in the form of its spherical coordinates Θ, Φ and R, as per the following diagram.

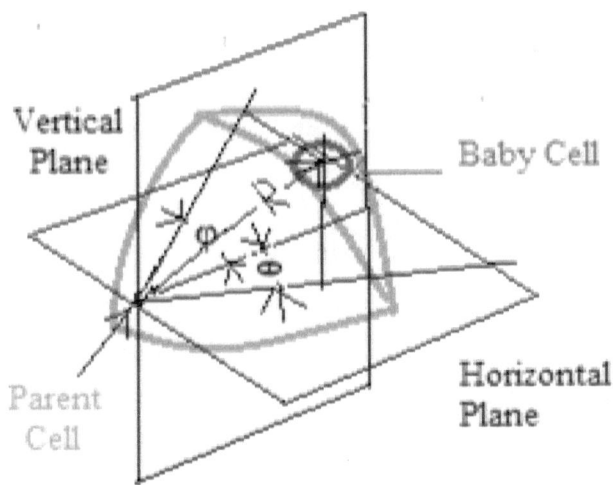

Spherical coordinates of the baby cells

Not only the cells produced by the zygote — no cell produced by even them or, any cells produced, thereafter could have produced any cells unless such information should have been furnished to all of them.

Don't you think, the biologists should have bothered to also know

- How each cell may come to know the information it is supposed to pass on to its baby cells?

- In what format each cell may be passing on such information to the cells produced by it? Obviously, it could have been only in some format other than the digital format known as binary system.

- Where such a program which has to be operational for the entire period of pregnancy may be residing — physically, in the brain of a pregnant woman?

Since a full-grown human fetus consists of about 38 trillion cells of more than 300 types, don't you think — after a few weeks of incubation, millions of cells must be undergoing the same type of conversion and such information should have reached each of them?

Do you think it may be possible for any earthling to develop such a "**cell-level information system**"?

Simply — impossible!

Is it not true that such a program could have been developed only by some such entity, the brain of which should have been thousands times more capable than the human brain?

Does it not lead us to conclude that at some stage, some such entity should have existed for which it should have been possible to develop such a system?

IF SO — IS THEREANYTHING WRONG IF WE MAY TREAT SUCH ENTITY AS "GOD"?

May be, I have been able to conceive that a fetus can't grow unless some such information system may have been developed by some entity only because I have worked on Management Information Systems for as long as about twenty years.

Well, this is not the only system some such entity should have been able to evolve.

IN THE NEXT CHAPTER, I AM GOING TO DESCRIBE ONE MORE SYSTEM OF THE SAME COMPLEXITY THAT COULD HAVE BEEN ALSO EVOLVED BY ONLY SOME SAME TYPE OF ENTITY.

CHAPTER - SIX

VISUALIZATION OF ANOTHER PROOF OF THE EXISTENCE OF GOD

Just think —if any such system that may be telling each cell of the yoke and the albumen to know "how it has to do — what it is supposed to do" to let the contents of an egg get converted into a chick inside the shell of an egg from the day, a hen lays an egg till it gets hatched, exists; how intelligent the entity for which it should have been possible to develop such system, should have been.

Although science has ascertained that the anatomy of an egg is of such type as shown in the following diagram.

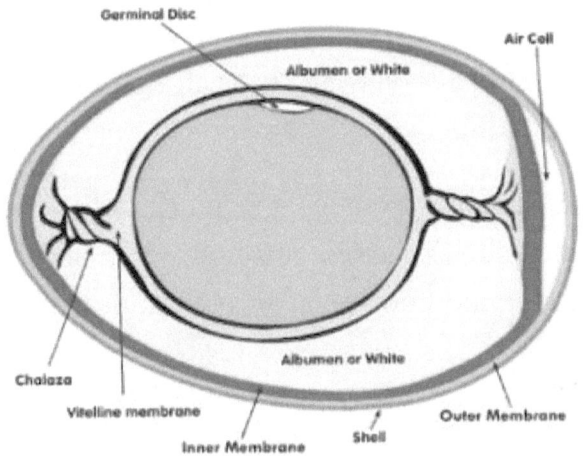

Is it not true that some entity should have calculated exactly which minerals have to be packed in how much quantity in an egg, which proteins have to be packed in how much quantity in an egg, how much fat has to be packed in an egg, how much water has to be packed in an egg and how much air has to be packed in an egg that may be necessary for the contents of an egg to get converted into a chick?

Just think, if so — how intelligent the entity for which it should have been possible not only to carry out such calculations;it should have been also possible to convert whatever a hen eats into such minerals in the requisite quantity, such proteins in the requisite quantity and such fat in the requisite quantity and enclose them in the shell of each egg?

You will get an answer to the query "Does any entity known as God exist?" only if you try to figure out which entity should have done all such calculations and to have made it possible.

Can we calculate exactly how much, which minerals, how much such proteins and how much such fat, how much water and how much air are required for creating a chick?

May we ever knowhow to produce all such minerals, all such proteins and the fat,that get packed inside the shell of each egg biologically and how to pack them in the form of the yoke and the albumen that get packed in the shell of each egg?

Obviously — not!

The following diagram shows the four stages of the growth of a chick inside the shell of an egg, as per https://comal.agrilife.org/files/2011/08/fromaneggtoachick_9.pdf.

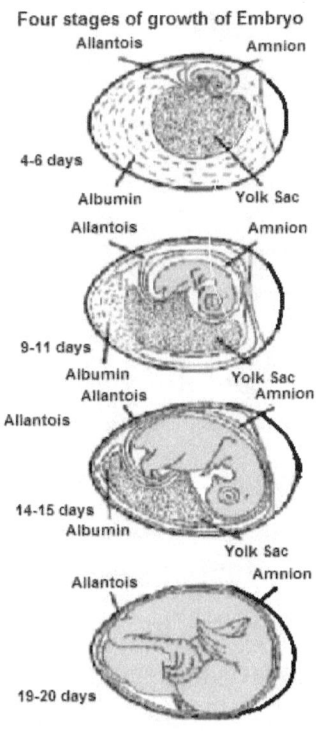

JUST THINK — WHICH ENTITY COULD HAVE DONE NOT ONLY ALL SUCH CALCULATIONS BUT ALSO MADE IT POSSIBLE TO CREATE SUCH MINERALS AND PROTEINS EXACTLY IN SUCH QUANTITIES AS NECESSARY AND PACK THEM IN EACH AND EVERY EGG?

Do you doubt — such an entity should have not only developed such a program, it should have also evolved some system of installing such program somewhere inside the shell of each egg?

DOES ANYBODY KNOW WHERE SUCH PROGRAM RESIDES IN AN EGG?

SHOULD WE MIND TREATING THE ENTITY FOR WHICH IT SHOULD HAVE BEEN POSSIBLE AS "GOD"?

If you are not sure about it — find some programmer in your office who may be able to write a computer program which may pass on the type of information the cells need to build the retina of an eye, the iris of an eye, the type of information the cells need to build the eyelids of the eyes and the type of information the cells need to build the heart and the type of information the cells need to build the brain and all other organs of a chick.

It has to be such type of program that may be running from the day an egg is laid by a hen till the day the baby chick comes out of the shell of an egg.

Until now, science is not even aware of the fact that some entity should have evolved some system of this type, though it should be very much — in place.

DOES IT NOT SATISFY YOU THAT IT IS WRONG TO DOUBT WHETHER GOD EXISTS OR NOT?

CHAPTER - SEVEN

THE WAY WE MAY PROVE THAT IT SHOULD BE POSSIBLE TO PRODUCE EVERY TYPE OF THING FROM NOTHING

Don't you think, though we don't know how to produce anything from nothing, it should not imply that it may not be possible to create every type of thing from nothing in the same way as it is possible to generate infinite unique numbers from zero in the following manner?

You don't have to go around the world to get convinced about such a possibility.

Simply look at the following equations.

$0 = +5 +15 -20$

$0 = +6 -40 -50 +84$

$0 = -30 -60 +90$

Are all numbers not unique?

If we may treat "zero" as "nothing" and treat the positive numbers as things which react with the baryonic matter and the negative numbers as the things which do not react with the baryonic matter — do such equations not serve as proof of the fact that it is wrong to think that it may not be possible to produce anything from nothing?

Actually, by such analogy — we have to treat even such numbers as "+8" and "−8" as unique numbers.

So we may treat even the following types of equations as a proof of the fact that it should be possible to produce every type of thing from nothing.

0 = +8 −8

0 = +500 −500

0 = +15000 −15000

Any doubt about it?

Just think — if it had not been possible to produce every type of thing from nothing, how the entire universe should have come into existence?

Since we don't doubt that nothing should have existed before the Big Bang, we are left with no choice other than the choice of believing that it should be possible to produce every type of thing from nothing.

I presume that you know — unlike the things which physically exist; things known as dark matter and dark energy are the type of things which do not react with the baryonic matter.

CHAPTER - EIGHT

RECENT EVIDENCE THAT SUPPORTS THE "THEORY OF REINCARNATION"

Though it is not certain whether every person gets reincarnated or not — itis certain that those who die in some accident or those who die an unnatural death get reincarnated.

Though even Dr Ian Stevenson has described the story of a Lankan girl in his book "Reincarnation and Biology" who died by sinking in a lake into which his mentally disturbed brother had thrown her — I have come to know of the following story of a woman who had died recently in a road accident in Rajasthan, which also confirms such belief and is, presently pursuing her MD in the USA.

THE STORY OF THE REINCARNATION OF A WOMAN WHO DIED IN A ROAD ACCIDENT IN RAJASTHAN

This evidence has been shared by Anand Dev Sharma on Quora.com. [1]

It relates to his friend Jinea who also belongs to the same place "Bhatinda", a well-known city of Punjab to which — he belongs.

As mentioned by him, she had raised a demand in December 2000, 4 months after her 3rd birthday, to visit Shri Ganganagar, Rajasthan — a town approximately 120 km away from Bhatinda.

Upon being asked the reason, she said, "My family lives there."

Everyone in her family was shocked when she told that her family lived elsewhere.

It all happened for days when, finally — her parents took her to a psychologist. When he directly conversed with Jinea, she told him about 120 acres of land, her cycle, the place where she used to keep her valuables and also some memories she made with her family.

Though it looked like a nightmare, the next morning the whole family went to that place and enquired about the family when the only son in the family opened the door.

Jinea hugged him tightly calling him "Veer ji" (which means — "brother", in Punjabi).

In the meantime, everybody came out of the house.

Her earlier father told them while she was sharing her memories with her then mother, that everything she had told was correct. He told her present father about the sudden death of their daughter in a road accident.

The very next day, Jinea told them about the accident in which she died and everyone there was dumb stuck.

She even told her last conversation with her brother which took place just before the accident.

Though she loved both families — she stayed with the present family.

For about four years, they kept on visiting the Ganganagar family once a fortnight. But after 4 years, her memory of past life started fading and after some time — she completely lost all of it.

However, the Ganganagar family not only accepted her as their daughter; they even promised to treat her as their daughter till the very end — including her wedding.

Does this instance not serve as adequate proof that something (generally referred to as our soul) works as a medium of carrying forward memories of the past life if anybody may die in some accident in such a manner in which she had died?

The most interesting thing about this instance is — it is not a unique case of this type.

You may visit the site:

http://reincarnationafterdeath.com/reincarnation-stories/ to know about eight more such stories to get convinced that the people who die unnaturally — invariably get reborn in the same way as in the case of Jinea or even have a look at:

http://theultimategoalofourlife.in/2017/09/20/there-is-a-long-time-gap-between-the-day-we-die-and-the-day-we-get-reborn-just-guess-what-the-souls-may-be-doing-all-along-during-this-time/

---

[1] https://www.quora.com/Does-anyone-believe-in-reincarnation/answer/Anand-Dev-Sharma

CHAPTER - NINE

PIECES OF EVIDENCE OF REINCARNATION – COLLECTED BY DR IAN STEVENSON

Though Dr Ian Stevenson (1918-2007) and his team did not leave any stone unturned to make the world aware of the same fact, the Indian researchers had made the world aware of during the Vedic period — science has not been able to come out of the stigma it had about the concept of reincarnation even though the criteria based on which it used to decide what is scientific and what is not scientific has undergone a sea change since the days of its infancy.

Though at the time of its inception, it had told the world that it was not scientific to believe even any such thing the existence of which can't be practically demonstrated may exist — it has turned turtle by having admitted that things known as "dark matter" and "dark energy" exist even though it has not been possible to practically demonstrate their existence.

The fact is — not only souls, even "dark matter" and "dark energy" are invisible.

Their existence has been justified only based on the fact that certain cosmic phenomena such as how the stars situated near the center of a galaxy may travel at more or less the same speed at which the stars situated near its outer edge travel, can't be explained unless we assume that things of the type of "dark matter" and "dark energy" exist.

But the craziest thing about it is — though it has assumed such an assumption to be valid, it has not been able to make up its mind about whether "*the fact that it is not possible to explain how anybody can tell anything related to his/her past life so precisely as many persons have been able to tell unless we assume that, if not all of us — some of us, get reincarnated*" may be valid or not.

DOES IT NOT AMOUNT TO HAVE DOUBLE STANDARD — ONE STANDARD FOR THE THINGS CALLED "DARK MATTER" AND "DARK ENERGY" AND A DIFFERENT STANDARD FOR THE THING KNOWN AS "SOULS"?

Though the work of Dr Stevenson and his team is like reinventing what had been already invented in India more than even thousand years back, it can't be denied that if science takes the data compiled by them by interviewing several children of several countries who could tell about the names of their parents, their brothers and sisters and their husbands or wives, the names of even their friends and neighbors during their past life — at its face value, science should acknowledge the existence of souls also in the same manner in which it has acknowledged the existence of "dark matter" and "dark energy".

Though even the story narrated by me in the previous chapter should have convinced you about the process of reincarnation, I am sure — the following story narrated by Dr Stevenson in his works, would further intensify your confidence in this process.

THE STORY OF THE REINCARNATION OF A LANKAN GIRL

One day, when a girl overheard her mother mentioning the name of an obscure town ("Katarangama") where she had never been during her present life informed her mother that she had drowned in a river when her "dumb" (mentally challenged) brother had pushed her into it and that she had a bald father named "Herath" — who sold flowers in a market near the Buddhist Stupa.

That she lived in a house that had a glass window in the roof (a skylight), where dogs in the backyard used to be tied up in a Hindu temple, outside of which people smashed coconuts on the ground.

Though this girl did get a few things wrong — 27 of the 30 such statements made by her panned out to be correct even though the two families had never met, nor did they have any mutual friends, co-workers, or other acquaintances in common.

She couldn't have been able to tell so many things so accurately unless we believe that when somebody dies, his or her memories may be getting whisked away by some entity.

Yes, I have in my mind the same entity that is known — worldwide, as "soul".

The Sri Lankan case is one of Stevenson's approximately 3000 such "past life cases" described by him in his 2,268-page, two-volume work called "Reincarnation and Biology" — which was published in the year 1997.

Though it is not known where the souls of the people who die, stay and keep the memory whisked away by them safely until it may be re-planted in the mind of some newborn baby; since such children use the words "I" and "my" in their narrations — the Hindus started using the term "rebirth" for such phenomenon.

With our computer knowledge, we may say that the souls may be taking a backup of the memory of the person in whom it should have resided, at the time of his/her death — the way, we take a backup of our files on a pen-drive.

Though it is difficult to ascertain — it is quite possible that they maybe uploading it on some type of website and maybe downloading it from there into the memory of their next host/hostess at the time of reincarnation.

The fact is — the existence of such things may be established through "Deductive Reasoning" only in the same manner in which the existence of "dark matter" has been established.[1]

---

[1] https://www.nationalgeographic.com/science/article/dark-matter

CHAPTER - TEN

THE TYPE OF FUNCTIONS ONLY SOULS MAY BE ABLE TO PERFORM

In his book **"The Power of Your Subconscious Mind"**, Dr Joseph Murphy has delved into certain Para-scientific capabilities of the subconscious mind at length.

One of such capabilities is the ability of the subconscious mind to transmit prayers to the subconscious mind of some other person even over inter-continental distances.

Before commenting on such a possibility — I would like to apprise you about an instance of this nature described by him on the pages 75 and 76 of the book.

I am reproducing it — verbatim.

"When you affirm health, harmony, and peace for yourself or another, and when you realize these are universal principles of your own being, you rearrange the negative patterns of your subconscious mind, based on your faith and understanding of that which you affirm.

"I chose to use this method when my sister was about to be operated for the removal of gallstones in a hospital in England. Her diagnosis was based on the usual hospital tests and X-ray procedures. She asked me to pray for her recovery. I was over six thousand miles away, but this did not disturb me. There is no time or space in the mind principle. Infinite mind or intelligence is present in its entirety at every point simultaneously.

"Several times in a day I withdrew all thought from the contemplation of my sister's symptoms and from the corporeal personality altogether. Calmly and comfortably, I affirmed as follows:

"This prayer is for my sister Catherine. She is relaxed and at peace, poised, balanced, serene, and calm. The healing intelligence of her subconscious mind that created her body is now transforming every cell, nerve, tissue, muscle, and bone of her being according to the perfect pattern of all organs lodged in her subconscious mind. Silently, quietly, all distorted thought patterns in her subconscious mind are removed and dissolved, and the vitality, wholeness, and beauty of the life principle are made manifest in every atom of her being. She is now open and receptive to the healing currents that are flowing through her like a river, restoring her to perfect health, harmony and peace. All distortions and ugly images are now washed away by the infinite ocean of love and peace flowing through her, and it is so.

"At the end of two weeks, my sister had another examination. Her X-rays were negative. Her doctor admitted that she showed remarkable healing and called off the scheduled surgery."

Just think, how such surgery could have been called off, that too — when his sister lived in a country far away from his country, on a different continent?

Though it is true that when we are in deep sleep during the night — our subconscious mind acts just like an ICU and starts rejuvenating our internal organs so that they may become fit to deal with the hassles of the next day — if I am correct, gallstones may be removed only "surgically".

Do you think, his prayers could have averted the surgical operation?

I doubt it very much.

But if we believe him, his prayers could have given her some relief only in the following manner.

The way, we can transmit our messages across various continents using the internet; if I may not be wrong — our souls may be also able to transmit such messages to the souls of other persons using some App of the same type as is used by us — employing some type of internet, similar to our internet.

So once his soul should have transmitted his prayers to the soul of his sister — it should have not been difficult for the subconscious mind of his sister to take the cudgels in its hands.

Should I tell you — why I think so?

An incident that had occurred in my life, makes me think so.

Once you know about the following incident that had happened in my own life — even you would be convinced that, somehow, souls are also able to communicate with the souls of other persons.

THE INCIDENT THAT MAKES ME TO THINK SO

In the year 1989, the Singhania group of Kanpur (a city close to the city where I live, which used to be an industrial hub of North India, once upon a time) had also not only started manufacturing scooters just like our factory, it had even advertised some vacancies.

Just like many other disgruntled employees of our company — I also responded against this advertisement.

Though I don't know why — when I reached Kanpur for the interview, I chose to go to the venue of the interview by public conveyance instead of hiring a cab which dropped me at a place a few kilometers away from the venue.

While I lookedfor a cab — I became nervous because it looked as though I would not be able to reach the venue of the interview at the time assigned to me for the interview.

Though my feet were stumbling — to my great surprise, a person approached me and asked me whether he could help me in any way.

When I told him about my problem — he offered to take me to the venue of the interview on his scooter.

On the way, when I asked him — how he happened to pass through the place where I was waiting for a cab, he told me that though he lived far away from that place; he used to come there only on weekends to collect payments of the goods supplied by him to his customers, but on that day some unknown soul woke him up and told him to reach this place to help somebody who is in trouble.

Don't you think — it could have not been possible unless my soul should have been able to send out an SOS message?

I hope it explains how the prayers of Dr Murphy should have been transmitted by his soul to the soul of his sister.

Though according to the criteria based on which it decides — what forms a part of mainstream science and does not form a part of mainstream science, such capability of our soul to transmit our messages to the souls of other persons should also become a part of mainstream science?

FOOT NOTE

This chapter is a modified version of the article "**A commentary on Para-scientific features of the subconscious mind**" published by me on medium.com on April 13, 2024. [1]

---

[1] https://subhashchandrasawhney.medium.com/a-commentary-on-para-scientific-features-of-the-subconscious-mind-363cfa791ba8

CHAPTER - ELEVEN

IT LOOKS AS THOUGH THE CURSES AND BLESSINGS GIVEN BY US ARE EXECUTED BY OUR SOULS ONLY

I don't know whether you know or not — the curses given by a harassed queen to the royal family of Mysuru of the Wodeyar dynasty had remained in force for as long as 405 years.

THE STORY OF THE CURSES GIVEN BY A HARASSED QUEEN TO THE ROYAL FAMILY OF MYSUSRU OF THE WODEYAR DYNASTY

It is well known that the royal family of the Wodeyar dynasty never had a natural heir since 1612 when they conquered the Srirangapatna region — Alemalemma, the wife of the outgoing king of that region is said to have run away with all the royal ornaments to Talakkadu, a small town along the Cauvery river.

THE REASON WHY THE QUEEN OF THE OUTGOING KING CURSED WODEYARS

Though the soldiers of Wodeyars chased the queen to get back the jewels; Alemalemma jumped into the river cursing the Wodeyars that "may Talakkadu be filled with sand and become a barren land and may Malangi (the stretch of the river where she had jumped into it) turn into a whirlpool" and "may the Wodeyars be never blessed with a natural heir".

She would have not cursed Wodeyars if their soldiers had not chased her to recover the ornaments.

THE WAY HER CURSES GOT EXECUTED

As a result of this curse, the Wodeyars did not get a natural heir until 2017.

Yaduveer Krishnadatta Chamaraja Wodeyar who is an adopted son of Rani Pramoda Devi and the 27th scion of the Wodeyar dynasty, got married to Trishika Kumari — the daughter of Harshavardhan Singh and Maheshwari Kumari, the royal couple of Dungarpur in June 2016.

It is a well-known fact that the royal family did not get a natural heir since then till December 6, 2017, when breaking the myth — a prince was born to them at 9.32 p.m. on December 6, 2017, in Bengaluru.

Yaduveer unveiled the following picture of the baby boy on his Facebook account.

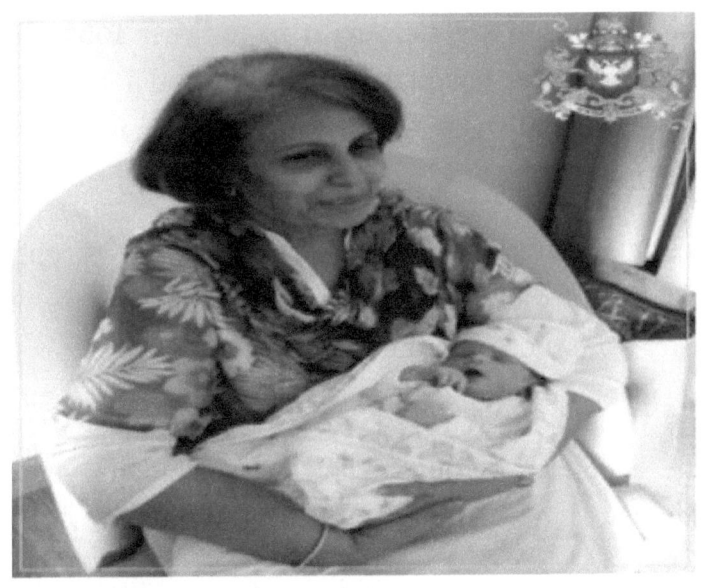

They hope that with the birth of this child, the 405-year-old curse will come to an end.

Though mainstream science may never be able to explain how such a curse should have been effective for as many as 405 years, I can tell — how curses work from my own experience.

If we look at how curse given by me to someone got executed, anybody can make out that only our souls execute the curses given by us to anybody.

THE REASON WHY I FELT LIKE CURSING THE CHIEF MEDICAL OFFICER OF THE FACTORY WHERE I WORKED AS AN ASSISTANT ENGINEER

I got married to a gynecologist in the year 1967. Since she worked in a hospital in Haryana and I worked in Tamilnadu, naturally — she resigned to join me at Tiruchirappalli.

Since this factory was located in South India, except for a handful of the North Indians — our colleagues were mostly from the South only.

Since there were hardly six or seven of us — we had made a coterie of our own which included even the Maharashtrian Chief Medical Officer of the factory.

It so happened that within a few months of our marriage, our company announced some vacancies for doctors in its hospital.

Since I had married a gynecologist — I told my wife to also apply against this advertisement.

Since the Chief Medical Officer of the company was a member of our group; I was damn sure that she would get selected against these vacancies more so because our company had a policy of giving preference to the dependants of the employees over the outsiders — to fill all vacancies.

But it came to me like a bolt from the sky when I learnt that she had not been selected.

Not just because she had not been selected but also because the Chief Medical Officer, on some pretext, went out of station on an official tour for a week or so, so that, he did not have to face the music.

When I got the whiff of it, I met the General Manager of the company and apprised him of the fact that since my wife had been working as a Gazetted Officer in a Government Hospital of Haryana, before marriage — there was no scope of any valid reason for her not having been selected against these vacancies.

I also told him that since she had resigned from her job after getting married to me — if she did not get selected it would create lots of problems for us.

Convinced by my argument — he took a second opinion from the Chief Superintendant of the Civil Hospital of the city and overruled the objection that had been raised by the Chief Medical Officer of our company.

But I got so annoyed over such a nasty stand taken by our Chief Medical Officer that I impulsively cursed him in my mind that may God punish him by letting him live amidst lepers.

If you may look at how this curse took shape — I am sure you, too, would agree that it could have not got executed in such an exquisite manner in which got executed unless it had been maneuvered by none else but by my soul only.

THE SYSTEMATIC MANNER IN WHICH THIS CURSE WAS EXECUTED – STEP BY STEP

Though I had described this sequence in which this curse got executed even in the chapter "**An episode that should startle anyone about the meticulousness with which the souls can execute a task**" of my book "**Not One — We have Three Domains of Science**", I am going to narrate it once again.

Somehow, it so happened that within one or two months of this incident — the Chief Medical Officer of the company conducted an operation of appendectomy on the wife of the General Manager even though it was not called for.

After having come to know that it was not at all necessary to conduct such an operation on his wife, the General Manager got so infuriated that he issued marching orders to the Chief Medical Officer asking him to look for some other job within three months.

However it never occurred to us that it would be so difficult for him to find a suitable placement somewhere since he was an FRCS, which was regarded as a very big qualification for the surgeons those days.

But destiny took its turn.

The fact is — he failed to find a job for himself even for as long as six months or so.

So he was compelled to tender his resignation.

Years rolled by and, one fine day, we got the news that he had joined a missionary hospital of lepers run by nuns — in a city about 100 km further down from our factory, where our group even went to congratulate him.

Thinking about the reason, why he rejected my wife, I may only say that maybe — because he did not like that if my wife also got a job, our take-home salary would exceed even his take-home salary.

Though this episode took place almost fifty-six years back, I kept on guessing — how any mistake committed by him led him to resign and then join a hospital for the treatment of leprosy as if it had been planned out by some human beings to let him get punished for having rejected my wife.

Don't you think — this curse should have been executed by some soul only to teach him a lesson of a kind?

But don't you think — if so, it could have been, at best, my soul?

Though we can't see the souls — since there is no other way, how we may explain how so many things should have happened in a row, one after another, in such a well-sequenced manner; we should not doubt whether souls exist or not.

Anyway, does it not imply that the curse given to the kings of the Wodeyar dynasty should have been also executed by the soul of the queen — none else?

Don't you think — there is no difference between "the waysuch curses got executed maybe explained only if we agree that some entities of the souls type exist" is akin to the manner in which science has figured out that "certain such phenomena as 'how the stars situated near the black holes of the galaxies may travel at more or less the same speed at which the stars situated near the edge of their galaxies travel' can't be explained as long as we don't assume that things of dark matter and dark energy type exist"?

If so, don't you agree that science should agree that there is nothing wrong in assuming that "souls may be also existing" much in the same way in which there is nothing wrong in assuming that "things known as dark matter and dark energy exist"?

CHAPTER - TWELVE

HOW SOULS MAY BE ASCERTAINING WHO HAS TO GET REINCARNATED WHEN AND WHERE

Mostly the people who believe in the theory of reincarnation have a notion that every gets reincarnated which does not appear to be a correct notion because not even one child out of ten thousand children is able to describe the things pertaining to their past life.

Since it is so, it becomes necessary to study — I doubt if anybody may have ever bothered to study

(i) What type of people get reincarnated.

(ii) How the souls of such persons may be ascertaining in which families they should get born on which date and at what time or, perhaps where they should get reborn on which dates and at what time.

(iii) Where the souls may be residing till their previous host/hostess get reborn?

(iv) How do they manage to keep the memories of the persons in which they had been living — intact till they download such memories in the mind of their next host/hostess?

Don't you think — human mind is incapable to conduct such study due to such reasons as I described in the ___

True, we can only guess.

Let us have a look at the way "**Theory of *karmas***" explains such things.

According to this theory, the people who may die in an accident or who may have died in an unnatural way, get reincarnated due to the following reasons.

(i) Since they could not perform all the tasks they were supposed to have performed if they would have not died in an accident or in an unnatural way — they get reborn to perform the tasks they could have performed only if they could have not died in such a manner.

(ii) Though all of us keep on earning fruits of the type of karmas, we keep on performing in our life — since we are not able to earn the fruits of all karmas performed by us during our lifetime even if we don't die in any accident or due to some unnatural way — such people get a chance of reaping the fruits of such karmas for which they may have not been able to reap the fruits of their karmas during their previous life cycle.

Again, though **"Theory of *karmas*"** tells why some people get reborn; it does not tell though some people get reborn within a decade (as in the case of the woman described by me in the eighth chapter of the book, some of them get born even after a gap of several centuries or even — a much longer gap.

The way we may guess why some people do get reborn within a decade and some people do not get reborn even after several centuries.

According to me, the souls may be calculating who has to get reborn when and where by using some type of astrology which may be much more accurate than the type of astrology being used by the astrologers of our planet.

They may be, astrologically, finding out the longitude and the latitude of such place as well as the time of birth where and when anybody born, may get predestined to reap the fruits of such karmas, the fruits of which — he/she should have not been able to enjoy during his/her previous life cycle.

OF COURSE, TO FIND OUT — EXACTLY WHAT TYPE OF ASTROLOGY THEY MAY BE USING, IS BEYOND THE SCOPE OF HUMAN MIND DUE TO THE REASON DESCRIBED BY ME IN THE SECOND CHAPTER OF THIS BOOK.

CHAPTER - THIRTEEN

THE WAY HINDU DEITIES PROVE THEIR EXISTENCE BY APPEARING IN THE FORM OF THEIR APPARITIONS

I would have not believed that Lt. Col. Russell Dean Martin may have seen an apparition of "Lord Shiva" on the battlefield fighting along with his soldiers against *Pathans* in Afghanistan in the year 1879, if I would have not seen an apparition of "Lord Yamaraj" in the year 1985 — myself.

In this chapter, I am going to describe both instances.

THE STORY OF LT. COL. RUSSELL DEAN MARTIN WHO HAD SEEN AN APPARITION OF LORD SHIVA IN THE YEAR 1879

While the fight was going on in Afghanistan, Martin used to send messages of his well-being regularly to his wife, who used to live in Agar Malwa of Madhya Pradesh state of India.

But after a few days, she got worried since she stopped receiving such messages from him.

However, one day while riding on her horse, she passed by the side of a temple of Lord Shiva known as the temple of "Baijnath Mahadev".

Fascinated by the sound of a conch and the mantras being recited inside the temple, she got down from the horse and went inside the temple where some Brahmins were worshipping Lord Shiva.

When the Brahmins noticed she looked very sad and tense, they asked her about the reason of her looking so sad and tense.

When she told them the reason — they advised her to perform "*Laghu-rudra Anushtthan*" by reciting the Mantra "*Om Namah Shivay*" for eleven days.

On their advice, she recited this Mantra for eleven days and decided that if her husband reaches home safely — she would get the temple renovated.

It so happened that on the eleventh day of the *Anushtthan*, a messenger came to give her a letter on which her husband had written, "I was regularly sending you messages from the battlefield but suddenly the *Pathans* surrounded us from all sides and we got trapped in a situation where there was no scope escaping death. Suddenly, I saw a Yogi of India with long hair carrying a weapon with three prongs (*Trishul*) fighting along with our soldiers.

"Seeing this man wearing a lion-skin fighting with his *Trishul* (Trident), the *Pathans* fled away.

"After having led us to victory, this man told me to do not worry. He had come there only to rescue me because he was pleased with your prayers."

After a few weeks, when Martin returned home — she took him to the temple of Mahadev to thank Lord Shiva, who had saved him.

It was then when he told her that — that man looked very much like the statue of Lord Shiva.

It is believed that after visiting the temple, they got the temple renovated and donated some money even for its maintenance.

Picture of Lt. Col. Martin and the temple they got renovated at Agar Malwa in Madhya Pradesh, in the year 1883

Though we don't know how far this story may be true, we have to believe it since it appears even on the website of the temple and the names of the British couple are engraved on a slab inside the temple premises.

THE FORM IN WHICH LORD YAMARAJ APPEARED BEFORE ME IN THE YEAR 1985

Though even the whole story of Martins is quite fascinating, you would be taken aback if I tell you that even I had seen an apparition of the Hindu deity known as *"Yamaraj"* — the God of Death in the Hindu pantheon.

Yamaraj —the God of Death

The fact is — on May 4, 1985 my younger brother had come to Lucknow (where I live) from Kanpur along with his wife to pick up our parents to take them with him to his place since my parents very much insisted that they shall feel more comfortable at Kanpur as his wife stayed all the time at home unlike my wife because we are a "working couple".

It was just when all of them sat in the car, I saw a white apparition of "*Yamraj*" — mounted on a white bull, with his mount standing just two feet behind the car, which stared at me and very clearly told me, "*I shall follow the car up to Kanpur and shall take your father with me from there.*"

It was a day, I shall never forget in my life since the very next day early in the morning at about 4 a.m., I got a call from my brother in a sobbing voice that he had to break a very sad news to me — the news that our father had breathed his last at about 1 a.m. in the night, the same day."

For several years I kept on wondering, "If *Yamaraj*could have come to our house, mounted on a bull to take away the soul of my father — if such Lord comes to collect the souls of the deceased people in India; He should come to collect the souls of the deceased people on a bull, in all other countries as well, where nobody has ever seen any such apparition in the same way as Martin seems have seen the apparition of

Lord Shiva even though he may have not even known whether any such deity exists in Hindu pantheon.

So far as I think — there was a reason why Lord Shiva had appeared in the form of an apparition.

People of other countries don't see any such apparition only because there is no such reason why they may see any apparition of any Hindu deity.

But, after hearing the story of Martins; it occurred to me that Yamaraj should have presented himself in the form of an apparition before me only so that I could recognize — who he was.

What I had seen, also tells that Lord Shiva should have also appeared in the form of an apparition wearing a lionskin and carrying a Trishul in his hand before Lt. Colonel Martin only so that he could, later on, know who he was.

It is not at all necessary that there may be, really anybody who may be wearing a lion skin, having a snake coiled around his neck, sporting a crescent moon in his headgear and having a river flowing out of his hair as all Hindus conceive but since the Hindus have been believing that Lord Shiva has such outfit — He should have appeared in such outfit even before some more people off and on; in the same way as Yamarajhad appeared before me on May 4, 1985 — mounted on a bull.

Who doesn't know that a river can't ever emerge from the hairs of anybody be it Lord Shiva or anybody else?

The fact is — the Hindu religion came into existence more than 10,000 years ago, which is such a long period over which nobody can guarantee whether the shape of the Hindu religion may be still the same as it may have been 10,000 years back.

Obviously, it should have undergone several changes over such a long time.

Maybe, originally, Lord Shiva may not have had Ganges flowing out of his hair but some people may have added this feature later on, for the reasons — best known only to them.

Though it appears funny to think that God may be feminine or masculine; these stories prove that just like us, somepeople may have seen the apparitions of even other deities — in the same way.

FOOT NOTE

This chapter is a modified version of the article "**No denying Lt. Col. Martin must have seen Lord**

Shiva as had been claimed by him" [1],which appeared on **medium.com** on December 9, 2020.

---

[1] https://medium.com/@subhashchandrasawhney/no-denying-lt-col-martin-must-have-actually-seen-lord-shiva-as-had-been-claimed-by-him-94274fec9d84

CHAPTER - FOURTEEN

A FACT THAT MAY DISAPPOINT THOSE WHO THINK – GOD MAY BE OMNIPOTENT

Although it may have been possible for only some superhuman entity to develop such information systems as have been described by me in the fifth and the sixth chapters of the book; it looks as though — it may not have become possible for it to prevent the colossal loss of the electromagnetic energy produced by all stars of the universe being incurred ever since the universe has come into existence because if it should have become possible, it would have done something to prevent such loss — long back.

It would have not mattered much if such loss would have not been of such magnitude as it happens to be.

AN ESTIMATE OF THE MAGNITUDE OF THE ENERGY BEING LOST PERENNIALLY

Has it ever occurred to you that hardly 0.0000001 percent of the electromagnetic energy produced by our star falls on the Earth and the rest of it goes to waste?

The fact is — what is true about our star, is true in the case of all other stars of the universe — also.

Surely, you shall be taken aback if I told you that almost 62.69×10^{60} KW of energy has already gone down the drain, by now.

Though I would like to take the credit for doing such calculations, I won't mind sharing the calculations with you.

Here are the calculations done by me.

THE CALCULATIONS DONE BY ME

Just look at the following diagram.

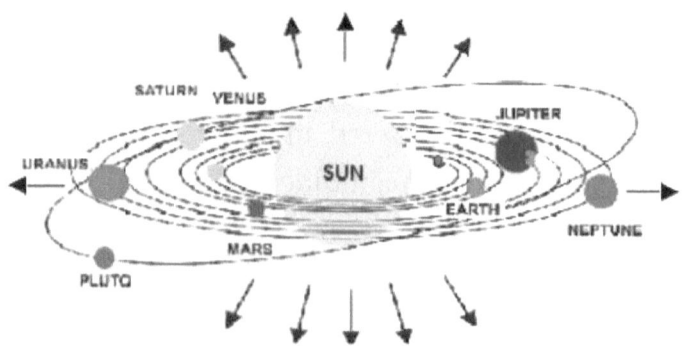

The diagram showing dispersal of the sunrays in 360° x 360° directions all around the Sun

Can you see what percentage of the sunrays produced by our Sun falls on the Earth and all other planets and their satellites?

It is only a miniscule.

Since our Sun generates 3.8×10^{26} W of energy per second, it works out to the generation of 11.98×10^{27} KW of energy per year.

After calculating how much energy is produced by the Sun per year, we may calculate what percentage of the energy produced by it falls on the Earth, based on this diagram.

As shown in this diagram, the rays produced by the Sun spread out in all directions.

Since the average distance of the Earth from the Sun is about 150 million km, the surface area of a celestial sphere of such radius works out to "$4 \pi \times 22500 \times 10^{12}$ sq km", which works out to 28.286×10^{16} sq km.

Likewise, since the average radius of the Earth is 6731 km, the surface area of the Earth exposed to the Sun is $2 \pi \times 45.3 \times 10^{6}$ sq km, which works out to 28.47×10^{7} sq km.

If we divide 28.47×10^{7} by 28.286×10^{16} and multiply the dividend by 100, we can calculate the percentage of the rays emitted by the Sun intercepted by the Earth.

This percentage works out to only 0.000,000,1 — which implies that hardly 0.000,000,1 per cent of the rays produced by the Sun fall on the Earth.

Just think — what percentage of the remaining 99.9999999% of the rays may be getting intercepted by other planets of the Sun and any other celestial bodies of the universe falling on their way, during their journey up to the boundary of the universe?

Let us assume that 2% of the power output of the Sun may be getting intercepted by the other planets of the Sun as well as all celestial bodies of the universe, that may intercept its rays on their journey up to the edge of the universe.

Though the power output of some stars of the universe is less than the power output of our star and the power output of some stars of the universe is more than the power output of our star — since the Sun happens to be a star of an average size in our galaxy, let us assume that it may be an average-sized star of even the entire universe and if so, we may also assume that the average power output per star may be the same as the power output of the Sun and calculate the power output of all stars that reaches the boundary of the universe per year, in the following manner.

Since there are 200 sextillion stars of various sizes, that is — 200×10^{21} stars in the whole space; the total power output of all stars per year may be calculated as follows.

The total power output of all stars per year = $(11.98 \times 10^{27}) \times (200 \times 10^{21})$, which works out to 23.96×10^{50} KW per year.

Let us assume that not only 2% of the rays produced by the Sun, get intercepted by the other planets of the Sun as well as on all the celestial bodies that may intercept them, during their journey up to the boundary of the universe — 2% of the energy emitted by all stars of the universe, maybe getting intercepted by various celestial bodies falling on the way of their rays also by the time they reach the boundary of the universe.

Based on this assumption, we may calculate the quantity of the energy of all stars of the universe reaching the boundary of the universe, as follows.

The quantity of the energy of all stars of the universe reaching the boundary of the universe per year = $0.98 \times 23.96 \times 10^{50}$, which works out to 23.48×10^{50} KW.

Since the age of the universe is about 26.7 billion years, we may calculate the total poweroutput of all stars put together, that have crossed the boundary of the universe since the evolution of the universe until now, as follows.

The energy emitted by all stars put together, that has reached the edge of the universe since the evolution of the universe until now = $23.48 \times 10^{50} \times 26.7 \times 10^9$, which works out to 62.69×10^{60} KW.

The worst part of it is — it will keep on getting lost for the whole life of the universe.

FOOTNOTE

I touched upon this topic not only in my book **"Existence of God"** which was published in the year 2016, I had posted such calculations even in the article **"Misconception about the Omnipotence of God"** on **medium.com**, on April 21, 2024.[1]

---

[1]
https://subhashchandrasawhney.medium.com/misconception-about-the-omnipotence-of-god-aeab101ea803

CHAPTER - FIFTEEN

THE REASON WHY WE DON'T KNOW SOME OF SUCH TECHNIQUES THAT WERE KNOWN TO SOME PEOPLE IN THE PAST

There are many such things some of which, I am going to recount in this chapter — science has not been able to tell how such things could have been possible even though such things are very much known to have become possible.

Though it does not seem to have occurred to anybody, I think — such things should have been explained in some manuscripts or another, that may have been archived in the six-storey library of Nalanda University but should have got burnt at the instance of either Bakhtiar Khiljee or the Brahmins to avenge the murder of Ashoka's grandson (as believed by Ram Puniyani, the president of "Centre for Study of Society and Secularism").[1]

According to me, we should have known the following things if this library had not been burnt.

1. HOW THE HINDU DEITY "HANUMAN" SHOULD HAVE BEEN ABLE TO FETCH A

LIFESAVING HERB FROM A PLACE WHICH IS 2327 KILOMETERS AWAY FROM CEYLON (NOW KNOWN AS SRI LANKA) ABOUT 7000 YEARS BACK – OVERNIGHT?

As shown in the following map, "Hanuman" had brought the lifesaving herb known as "Sanjivani" from the Himalayas which is as far away as 2,327 km from Sri Lanka to save the life of Laxman, the younger brother of Lord Rama about 7,000 years back — who was lying on the deathbed.

Distance of Himalayas from Sri Lanka

The fact is — it could have been a myth only if the sage Valmiki, who had attended the coronation of Lord Rama, would have not written in his Ramayana that he had fetched not only such herb from the Himalayas overnight but he had brought even the patch of the rock over which it grew — in such a manner as has been depicted in the following diagram.

We could have doubted such possibility only if Laxman, who had lost consciousness while fighting against the warriors of Ravana, the king of Lanka, would have not been revived by some physician of Ceylon — by administering a decoction of this herb to him.

Though the speed at which he should have flown in the air to fetch this herb may be calculated in such a manner as has been explained by me in the article **"The way it should have become possible for Hanuman to fly at a speed of 517 km/hour"**[2] — which had appeared on **medium.com** on November 26, 2023, the fact is — we don't know how he could have been able to fly in the air at such a high speed.

But I am pretty sure that the technique used by him should have been described in some manuscript or another of this library.

Even I wondered — how it could have been possible until I did not watch the following video which shows a ceremony that is celebrated in India every year on "Jaljhulani Ekadashi" (according to the Vikrami calendar) in the village "Hatpiliya" in the "Dewas" district of Madhya Pradesh at the temple of Lord Vishnu's fourth incarnation "Narasimha".

VID-20240724-WA0022 (7).mp4

On this day, the priest of the temple takes out the stone idol of the "Narasimha" to bathe it in the nearby river "Bhamori".

After completing the routine devotional ceremony, the priest drops this idol weighing about seven and a half kilograms into this river for bathing it.

Amazingly, instead of sinking, this idol flows back in the opposite direction and comes back to the priest, not just once — again and again.

Thousands of devotees from all over the country come here to see this wonderful scene.

2. HOW THE EGYPTIANS SHOULD HAVE BEEN ABLE TO CONSTRUCT THE PYRAMIDS WAY BACK AROUND 2550 YEARS BC?

Do you think — it would have become possible for the Egyptians to build the gigantic pyramids (of which the pyramid of Giza is as high as 417 meters) as they could build if they had not known the technique of hauling the type of stones they used from somewhere far away from where they should have quarried them since there was no such mountain in Egypt from where they could have quarried them and the technique of placing them so precisely one above another, each of which weighs as much as about 2.5 tons on an average?

The Giza Pyramid

*We should not doubt that the technique employed by them should have been described in some manuscript of this library since a similar technique should have been used even in India by the Indianssince they had been able to build the 7.77 meters-high canopy of the dome of the temple of Tanjaur of Tamilnad(built by a Chola emperor), which weighs as much as "**80 tons**"— in the year 1010 CE (or 1193 CE) by bringing the stone from some distant quarry only since there is no such mountain close to Tanjaur from where such stone should have been quarried.*

The only difference is that — though the whole world knows about the pyramids, even many Indians don't know about it.

3. HOW IT COULD HAVE BECOME POSSIBLE FOR SHAKUNTALA DEVI TO CALCULATE THINGS FASTER THAN EVEN THE UNIVAC COMPUTER OF SOUTHERN METHODIST UNIVERSITY, DALLAS USA?

Whether you know or not, the fact is that — the Indian lady Shakuntala Devi (1929-2013) was able to not only multiply such large numbers as 7,686,369,774,870 and 2,465,099,745,779 correctly just in 28 seconds in the year 1980, in the UK; though we can't calculate even the square root of a 5 digit number mentally in 50 seconds, she had been able to calculate even the 23rd root of the following 201 digit long number just in 50 seconds even though the Univac 1101 computer took 62 seconds to calculate the same root of the same number when she had been tested at Southern Methodist University, Dallas, USA, in 1977.[3]

916, 748, 676, 920, 039, 158, 098, 660, 927, 585, 380, 162, 483, 106, 680, 144, 308, 622, 407, 126, 516, 427, 934, 657, 040, 867, 096, 593, 279, 205, 767, 480, 806, 790, 022, 783, 016, 354, 924, 852, 380, 335, 745, 316, 935, 111, 903, 596, 577, 547, 340, 075, 681, 688, 305, 620, 821, 016, 129, 132, 845, 564, 805, 780, 158, 806, 771

Shakuntala Devi

The most amazing thing about her ability was — she never used any paper or pencil to do any such calculations.

Anyway, let us look at all the possible ways of finding nth root of any number "A" according to our present knowledge. [4]

We don't know of any way of calculating roots of any number other than the "Logarithmic method" and the "**Newton-Raphson method**".

Since she couldn't have mugged up the logarithmic tables, we may say — she couldn't have used logarithmic tables to calculate the 23rd root of this number.

This leaves us with the option of calculating the nth root of such number only through the "**Newton-Raphson method**".

I am going to explain how we may use this method to calculate the "nth root" of any number "A".

To calculate the "nth root" of any number by this method, first of all, we have to guess — what may be the value of the nth root.

Let us denote the first guess as "x_1".

Then we have to calculate the first approximate value "x_2" of the nth root, as follows for $k = 1$, k being the count of iteration.

$$x_{(k+1)} = [(n-1) x_k + A / x_k^{(n-1)}] / n$$

$$\Delta x_k = [A / x_k^{(n-1)} - x_k]$$

Then we have to go on increasing the value of k by 1 every time to recalculate the value of $x_{(k+1)}$ till we get $|\Delta x_k| = 0$.

Now, let us assume that she should have iterated such calculations 20 times.

In that case, it would have been necessary for her to keep in her memory the 201 digit number, the values of x_k and the values of Δx_k for $k = 1$ to 20 and keep on deleting the values of $x_{(k-1)}$ and $\Delta x_{(k-1)}$ from her memory, after every iteration.

So she should have taken only 2.5 seconds per iteration.

As such it could have not been at all, possible for her to do so many iterations mentally at a speed of 2.5 seconds per iteration.

Just think — if so, what does it lead us to assume?

The main reason why science has not been able to solve such riddles

If I am not wrong — since it is a fact that pyramids have been built and she was able to solve such mathematical calculations at computer-like speeds, though we can't say that to find out how it should have become possible, we can't say that to find out — how it could have been possible does not form a part of mainstream science; science should have , now agree that mainstream science may have two shells — the "inner shell" and the "outer shell".

Though these investigations form a part of the "inner shell" — science is working on only the "outer shell" all the time.

We may assume that she may have used either some third technique or some technique of remembering so many intermediary constants as should have been necessary if she should have used the **"Newton-Raphson method**" *which may have been described in some manuscript kept in the library of Nalanda university of which she may have not only been a student or some faculty; she may have been a reincarnation of such student or such faculty.*

Of course, it could have been possible only if she had managed to let such a technique not get wiped out of her memory since almost everybody forgets what he or she may have remembered about their past life — by and by

---

[1] https://m.thewire.in/article/history/did-bakhtiar-khilji-destroy-nalanda-university

[2]https://subhashchandrasawhney.medium.com/the-way-it-should-have-become-possible-for-hanuman-to-fly-at-a-speed-of-517-km-hour-094b196cdafe

[3] https://petersmagnusson.org/2016/01/15/shakuntala-devi-the-human-computer/

[4] https://www.quora.com/How-did-Shakuntala-Devi-perform-such-huge-calculations-with-great-speed-and-precision/answer/Subhash-Chandra-Sawhney

CHAPTER - SIXTEEN

PARA SCIENCE RELATED TO THE EPISTLE WRITTEN BY PAUL IN CHRISTIAN NEW TESTAMENT

I am going to talk about the Para Science related to the epistle "Be not deceived; God is not mocked: for whatsoever a man soweth, that shall he also reap" written in the Christian New Testament through what is supposed to have saved Donald Trump during the attempt of assassination on July 13, 2024in Las Vegas, Nevada at his campaign rally.

Trump, himself, said that he would have not got saved if he would have not tilted his head exactly to such an extent to which, he had tilted his head exactly when he had tilted his head.

A BRIEF ON THE TYPE OF KARMAS DONALD TRUMP HAD PERFORMED 48 YEARS AGO

48 years ago, when the International Society for Krishna Consciousness (ISKCON) was planning to organize the first Rath Yatra (Procesion of the chariot of Lord Jagannath) in New York City, challenges were galore.

While the grant of parade permit at the Fifth Avenue was nothing short of a miracle, finding a huge empty site where chariots could be built was also never going to be easy.

They knocked at the doors of every person possible, but in vain. It was then that Trump emerged as a ray of hope for the Krishna devotees. With ISKCON celebrating its 10th birthday in 1976, devotees in NYC were planning the first Rath Yatra there.

Exactly 48 years ago, in July 1976, Donald Trump had helped ISKCON devotees organize Rathyatra ceremony by providing his train yard for the construction of the chariots for free. The first chariot procession of Jagannath known as the Lord of the Universe by Hindus had kicked off on the streets of NYC in 1976, with the assistance from the then 30-year-old emerging real-estate mogul in United States of America - Donald Trump.

They had permission to use the fifth Avenue, which in fact was a big deal.But they needed an empty site close to the starting point of the parade route, to build the massive wooden carts. everyone they asked said "No". They were concerned about insurance risks which was understandable.Desperation of devotees had reached peak, hopes were nearly shattered. Almost all the film owners who were approached reportedly said they were in the process of selling the property at the Pennsylvania rail yard, which was marked as the perfect location for cart making. A few days later, someone told them that Donald Trump had purchased the old yard. But still there were concerns as a dozen other landlords they had already asked said "No", and "Why would Trump be different?"

Nevertheless, the devotees went to his office with a big basket of Maha Prasadam and a presentation package. His secretary took it but warned the devotees, "He never agrees to this kind of thing. You can ask but his answer is going to be "No". Have faith in Mahaprabhu they say, and the miracle was bound to happen!

Three days later, Trump's secretary called up the devotees saying ,"I don't know what happened but he read your letter, took a bit of the food you left, and immediately said "Sure, why not?".

The secretary then said "Come on down and get his signed letter of permission." Yes, Trump had signed the papers giving permission to use the open rail yards for the construction of Rath Yatra carts.

Receiving approval for a parade permit was another daunting task that the devotees faced.

On behalf of the Hare Krishna Movement, Tosan Krishna Das, who was incharge of getting the permit had submitted a written proposal to the authorities. The Police department had initially said "Yes" but they again said "No" as there was a mayor's edict since 1962 against allowing new parades on Fifth Avenue. With no option left, Tosan finally approached the chief of Police in Manhattan. Nobody had any clue about how the chief would respond to the application. But after a careful review, the police chief signed the papers with a smile though he said. "I don't know why I am doing this".

And then — he signed it.

Like other corporate company owners, Trump also could have easily rejected the proposal.

The devotees thought — Police Chief was no exception either.

Why they did not say "No" is still a question unanswered, as devotees cite it as a blessing of Lord Jagannath.

Though Donald Trump had also felt that only some divine power should have his head at the right moment to the right extent only, when he was campaigning at his rally in Pennsylvania — the attack on Trump and his survival, coinciding with the Jagannath Ratha Yatra celebrations, has led many ISKCON devotees to see this as an instance of divine intervention, bringing back memories of Trump's connection with the way he had helped them to organize the first Rath Yatra of the lord Jagannath in NYC.

But do you agree— all suchthings are covered by Para Science only — not by Mainstream Science?

DISCLAIMER

This book is intended to provide information and insights about the limitations of the human mind based on the author's research, experiences, and perspectives. It is not intended as a substitute for professional advice, diagnosis, or treatment from a qualified mental health professional, psychologist, or medical provider.

The content in this book should not be used to diagnose or treat any mental health condition or to make any assumptions about one's cognitive abilities or limitations. Readers should seek the advice of a licensed professional for any concerns related to their mental health or cognitive functioning.

The author and publisher disclaim any liability or responsibility for any loss, damage, or injury caused or alleged to be caused directly or indirectly by the use of the information contained in this book.

Reader discretion is advised, and readers are encouraged to use their own judgment and seek professional advice where necessary.

PROFILE OF THE AUTHOR

Former Senior Manager (Management Services) at Scooters India Limited, Lucknow and the Founder Chairman of the Lucknow Chapter of Computer Society of India, he has a B.Sc. (Engg.) HONS. from Thapar Institute of Engineering & Technology.

He lives in Lucknow (India).

His other books are:

1. Productivity Management: Concepts and Techniques

2. Taking a Lesson from the Mistakes of Others

3. Existence of God

4. The Mathematics of Uncertainties

5. Not One — We have Three Domains of Science

6. Revolutionizing Science: Discovering the Potential of Untapped Territory of Science

While his website is "theultimategoalofourlife.in", his e-mail ID is "subhashchandrasawhney@gmail.com".

MAY I ASK YOU FOR A SMALL FAVOR?

First, I want to say a big thanks for reading this book. You could have chosen any other book, but you took mine, and I appreciate this.

I hope you have at least a few actionable insights that will positively impact your daily life.

Can I ask for 30 seconds more of your time?

I'd love it if you could leave a review of the book. That will help me grow my readership by encouraging folks to take a chance on my books.

It will take less than a minute of your time but will tremendously help me reach out to more people.

If you liked this book, please consider posting an honest review where you got this book from. And I'd love to see your review. Thanks for your support.

FRONT COVERS OF THE OTHER BOOKS OF THE SAME AUTHOR

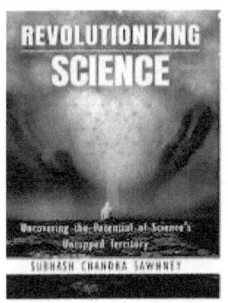

www.ingramcontent.com/pod-product-compliance
Lightning Source LLC
Chambersburg PA
CBHW020434220526
45464CB00002B/701